GUIDELINE FOR DRUG COURTS
ON SCREENING AND ASSESSMENT

Roger H. Peters[1] and Elizabeth Peyton[2]

*Prepared for the American University, Justice Programs Office, in association
with the U.S. Department of Justice, Office of Justice Programs,
Drug Courts Program Office*

May 1998

---

[1] Associate Professor, Department of Mental Health Law and Policy, the Louis de la Parte Florida Mental Health Institute, University of South Florida.

[2] Executive Director, National TASC — Treatment Accountability for Safer Communities.

Acknowledgments

The authors express sincere appreciation to the following persons who volunteered their valuable time and expertise to the development of this document:

Hon. William G. Meyer
Denver Drug Court Project
Denver, CO

Marc Pearce
National Association of Drug
  Court Professionals
Alexandria, VA

Martin O. Conoley
Santa Barbara County Probation
Santa Barbara, CA

Ed Brekke
Los Angeles Superior Court
Los Angeles, CA

Caroline Cooper
The American University
Washington, DC

Jody Forman
Charlottesville, VA

Hon. Richard S. Gebelein
Delaware Superior Court
Wilmington, DE

Joseph Carloni
Pathway Treatment Services
Pensacola, FL

Valerie Moore
IN ACT, Inc.
Portland, OR

Table of Contents

FOREWORD 1

INTRODUCTION 3

    What are the differences between screening and assessment? 3

    Goals of screening and assessment. 4

    Characteristics of screening and assessment 5

    What factors help to shape the drug court screening and
    assessment process? 6

    Importance of drug court screening and assessment 6

    Performance benchmarks for drug court screening and assessment 8

DRUG COURT SCREENING 9

    Who should conduct screening? 9

    Steps in conducting screening 9

    What information should be included in a drug court screening? 11

    Core screening elements 11

    Screening issues for women 12

    Screening for mental health problems 13

    Screening for suicide 14

    Screening for motivation and readiness for treatment 14

    Use of self-report information 15

    What instruments should be used in drug court screening? 16

    Key issues in selecting screening instruments 17

    Substance abuse screening instruments 17

Mental health screening instruments 18

Motivational screening instruments 18

What screening information is most relevant to the court? 18

DRUG COURT ASSESSMENT 21

When should drug court assessment be conducted? 21

Who should conduct assessment? 21

What information should be included in a drug court assessment? 22

Areas for detailed assessment 22

What instruments are available for assessment of
participants in drug court programs? 24

What assessment information is most relevant to the court? 25

Obtaining release of confidential information 25

SUMMARY 27

REFERENCES 29

OTHER RELATED RESOURCE MATERIALS 33

APPENDIX A: Selected Instruments 35

APPENDIX B: Availability and Cost of Screening Instruments 43

APPENDIX C: Addiction Severity Index 47

APPENDIX D: Single State Alcohol and Drug Agency Directors 63

Foreword

This guideline is written to help drug courts develop effective policies, procedures, and techniques for screening and assessing treatment needs of drug court participants. This document describes the principles and methods of screening and assessment of adult drug court participants, and gives drug courts specific tools and information to establish and sustain screening and assessment processes. While much of the information here will be helpful for juvenile drug courts, specific guidance for juveniles should be obtained from other sources.[3]

Several key principles and strategies for conducting effective screening and assessment described in the guideline are derived from experiences of existing drug courts and other community-based substance abuse treatment programs for offenders. Several useful guidelines and monographs on screening and assessment for criminal justice and non-criminal justice populations are included as references in the back of this guideline. Readers interested in additional information on this subject are encouraged to use these resources. This publication is one of several technical assistance monographs for drug courts that the U.S. Department of Justice, Office of Justice Programs, has developed.

This document presents general issues related to screening and assessing drug court clients, describes the processes and elements of screening and assessment in detail, summarizes key issues for drug courts to consider as they screen and assess participants, and provides resource materials for those seeking additional information.

---

[3]See for example Center for Substance Abuse Treatment (1993) *Screening and Assessment of Alcohol- and Other Drug-Abusing Adolescents.* Treatment Improvement Protocol Series, #3. Rockville, MD.

Introduction

What Are the Differences Between Screening and Assessment?

Screening and assessment are often described as discrete events completed by using specific instruments. In fact, screening and assessment are part of an ongoing decision-making process that examines information on substance abuse and criminal history, motivation for treatment, educational and employment factors, and other problem areas. Information gathered during screening and assessment is used to develop a treatment plan that will be updated over time to reflect participant progress, significant life events (e.g., relapse, changes in living arrangements), and changing service needs. While use of structured instruments is a core element of screening and assessment, this activity must be supplemented by an individual interview, review of archival materials (e.g., criminal justice records, treatment records, drug test results, employment records), clinical observation, and discussions with probation officers, family members, or significant others.

Although part of a continuous process, *screening* determines eligibility and appropriateness for participation in drug court, while *assessment* helps to identify specific types of services and determine the intensity of treatment needed. Screening is conducted in the very early stages of drug court involvement and typically precedes assessment and other diagnostic activities. Drug court screening typically consists of two steps: (1) *justice system* screening to decide if the prospective participant meets predetermined eligibility requirements related to criminal history, offense type and severity, etc.; and (2) *clinical screening* to determine if the prospective participant has a substance abuse problem that can be addressed by available treatment services, and if there are other clinical features (e.g., serious mental health disorders)

that would interfere with an individual's involvement in treatment. Once the initial screening decision is made, assessment helps to determine which types of services should be provided, and in what sequence these services should be provided. Diagnosis is part of the more detailed assessment process, and summarizes the pattern of current symptoms and functional impairment for several types of disorders (e.g., substance use disorders, mental health disorders).[4]

Several different drug court professionals such as prosecutors, public defenders, treatment staff, probation officers, court administrators, and pretrial services/TASC (Treatment Accountability for Safer Communities) staff, are often involved in screening.[5] A nationwide survey of drug court programs (Cooper, 1997) found that, for pretrial drug court programs, initial justice system screening is usually conducted by the prosecutor and either pretrial services or other drug court staff. Justice system screening is usually conducted by the prosecutor and probation officer in post-conviction programs. In most drug courts, the judge and prosecutor provide the final review of program eligibility, although the defense counsel is also involved in identifying and screening eligible cases. Once justice system screening is completed, a clinical screening is provided. In 38 percent of drug courts surveyed (Cooper, 1997), more than one agency is used to conduct clinical screening. These agencies include the drug court program, probation, private treatment agencies, the county health department, pretrial services, and TASC.

---

[4]This is typically accomplished through the use of the Diagnostic and Statistical Manual of Mental Disorders, Fourth Edition (DSM-IV), and specialized diagnostic instruments.

[5]Several jurisdictions, including San Diego, New Haven, Seattle, and Jacksonville use police as a referral source to drug court, and police officers are part of the screening process.

Staff who provide screening may not have extensive experience in assessment, diagnosis, and treatment issues. In contrast, assessment is typically conducted by substance abuse treatment professionals who have specialized education and training in these areas. While justice system and clinical screenings are usually completed in 5-30 minutes, assessment requires at least 1-2 hours. Assessment is more comprehensive in scope and provides much more detailed information, including examination of specialized areas such as diagnosis of mental health disorders. Assessment and related diagnostic information contribute directly to developing an individualized drug court treatment plan. The treatment plan for each participant enables the drug court to track problem areas, services provided, and progress toward program completion.

Some drug courts provide assessment instead of an initial screening. Although this approach provides more comprehensive information to guide initial placement in different types of services (e.g., residential versus outpatient), it is very time-consuming to provide a full assessment for all potential participants. Potential participants are also more likely to provide accurate self-disclosure of assessment information after they have been admitted to the drug court program. As an alternative to providing a full assessment at the time of initial screening, drug court programs may choose to implement a brief screening process. Many other drug courts have found that clinical screening before full admission to a drug court program serves several important functions, and screening is also an important function of many pretrial services, TASC, and other client management programs.

Goals of Screening and Assessment

The goals of screening are to:

- Determine if legal and statutory eligibility requirements are met;

- Determine the presence of substance use, mental health disorders, and medical conditions, including infectious disease;

- Define major areas of strengths and deficits;

- Determine if the severity of substance abuse problems is appropriate to the level of available drug court services;

- Weed out persons who do not have substance abuse problems;

- Identify individuals with a history of violent offenses/behavior;

- Identify environmental factors (e.g., employment, residential stability, relationship issues) or other disorders (e.g., mental health problems, cognitive deficits) that may undermine the individual's involvement in the drug court program or create an unacceptable public safety risk;

- Identify minimum level of security or supervision needed to promote public safety;

- Identify motivation, including perceived benefits and disadvantages of participation in the drug court program;

- Orient the potential client to program requirements; and

- Obtain consents for records and access to collateral contacts.

The goals of assessment are to:

- Examine the scope and nature of substance abuse problems;

- Identify the specific psychosocial problems to be addressed in treatment, including mental health disorders;

- Understand the impact substance abuse has had on the individual, including its influence on criminal involvement;

- Identify specific physical problems to be addressed in treatment planning;

- Identify the full range of service needs, pursuant to treatment planning;

- Match participants to appropriate types of drug court services; and

- Identify specific employment and educational deficits.

Characteristics of Screening and Assessment

The following chart summarizes issues to consider in developing drug court screening and assessment activities, and illustrates several differences between these activities.

	Purpose	Key Components	By Whom	Time and Cost Considerations
Legal Screening	To determine legal eligibility To examine public safety risk	• Current charge • Criminal history • Circumstances of offense	Criminal Justice System • Prosecution • Defense • Probation • Pretrial Services • TASC • Court • Police	These activities are conducted under normal criminal proceedings; cost is minimal.
Clinical Screening	To determine appropriateness of treatment and the individual's willingness and readiness for treatment	• Program explained • Releases signed • Brief assessment of substance use, social history, other disorders • Motivation/ willingness to participate	• Drug court case manager • Pretrial Services • Probation • TASC • Treatment Provider	Typically 5-30 minutes. Costs are associated with instruments, staff time, and staff training
Clinical Assess-ment	Diagnosis, admission and treatment planning	• Examine scope and nature of substance abuse problem • Identify full range of service needs, pursuant to treatment planning • Match participants to appropriate services	• Clinically trained and qualified counselor, psychologist, psychiatrist, social worker, nurse	1-2 hours or more, depending on the nature of problems. Costs are associated with instruments, staff time, and staff training

What Factors Help to Shape the Drug Court Screening and Assessment Process?

Several program features often determine the scope and context of screening and assessment activities. They are:

- **Treatment options** that are available to the drug court program. If several options for treatment services exist, more information is needed to guide participant placement.

- **Number of referrals** to the program. If the number of potential participants is large compared with the number of participants that actually enter the program, a separate screening process is a cost-effective way to "screen out" inappropriate candidates and focus resources on eligible participants.

- **Qualifications of screening staff.** Screening can often be conducted by staff members who do not have extensive clinical training. If non-clinical staff make initial eligibility decisions, a brief screen will help them identify appropriate participants, while clinical assessments are conducted by professional treatment staff.

- **Eligibility criteria.** If the drug court accepts all persons who meet legal requirements whatever their level of involvement with substance abuse, then clinical screening might be an unnecessary step. However, it is important to note that although this approach expedites the drug court admissions process, it may ultimately waste time by processing individuals who may not benefit from treatment interventions.

- **Client placement criteria** in a particular jurisdiction. There has been a recent movement to institute guidelines for referral and placement for various levels and durations of treatment. These "patient placement criteria" are influenced by the need to ration services and to standardize the referral and placement process. This movement is being driven by managed care, a variety of strategies that many states are using to allocate Medicaid and other funding for behavioral health (substance abuse and mental health) and other health care services. One widely used example of patient placement criteria has been developed by the American Society of Addiction Medicine (ASAM, 1996).[6]

Note: The above features should be examined by the drug court management team when developing policies and procedures related to screening and assessment, and should be reviewed periodically.

Importance of Drug Court Screening and Assessment

Candidates for drug court programs typically have a wide range of substance abuse, mental health, and other health-related disorders, in addition to many psychosocial problems related to employment and financial support, housing, family and other social relationships, transportation, and unresolved legal issues such as child custody. Many of these deficits are not clearly apparent through examination of criminal justice records alone, but can be revealed through an individual interview, drug testing, and use of specialized instruments. The rates of substance abuse disorders, mental health and personality disorders, suicidal behavior, physical and sexual abuse, and other health-related disorders such as TB and AIDS are much higher among criminal justice populations than among general community samples (National GAINS Center, 1997; Peters and Bartoi, 1997), and often go undetected in criminal justice settings. (Teplin and Schwartz, 1989). Non-detection of these disorders often leads to:

---

[6]Similar placement criteria have been adapted for use with criminal justice populations, including placement criteria developed by the Colorado Judicial Department, which are in use by the Colorado Departments of Correction, Probation and Community Corrections.

- Misdiagnosis;

- Neglect of appropriate interventions;

- Inappropriate treatment planning and referral;

- Over- or under-treatment of mental health symptoms with medications;

- Disruption of treatment services and demoralization of other participants; and

- Poor treatment outcomes (Drake et al., 1993; Peters and Bartoi, 1997).

An effective screening and assessment system helps to integrate this diverse information to form a comprehensive picture of each individual participant. Such integrated screening and assessment approaches are associated with more favorable treatment outcomes among individuals with multiple problem areas (Kofoed et al., 1986). An integrated screening and assessment system provides an important foundation that supports other drug court functions, including treatment planning, placement in treatment, and identification of the need for ancillary services. Screening and assessment marks the beginning of the drug court process and provides the core information needed to identify prospective drug court participants, evaluate their eligibility and appropriateness for participation, and begin the process of applying the services and sanctions that characterize drug courts. Information gathered during screening and assessment provides the basis for productive involvement of participants in the drug court program.

Participants in drug court programs and other substance abuse programs have reported that screening and intake interviews are among the most important of all treatment services that they receive. Screening and assessment activities provide a structured way for the court and the treatment provider to become familiar with the participant, and for the participant to become familiar with the goals and expectations of the program. Screening and assessment provide an important opportunity to develop motivation and commitment to the drug court program. This development is accomplished through the sensitivity of the drug court screener/assessor in expressing concern for and understanding of the psychosocial problems that have developed over time, discussing the relationship between substance abuse-related problems and recent criminal activity, eliciting the individual's goals (e.g., sobriety, employment), and emphasizing the benefits that can be achieved through participation in the drug court program (e.g., reunification with children, vocational training, avoidance of incarceration or criminal record).

Information gathered during the screening and assessment process describes the unique characteristics of each participant. It forms the basis for personal interaction with drug court staff, enables decision makers to place the participant in the most appropriate program available, and enables staff to determine if additional supports and services are needed to promote the participant's progress and success. In addition, the information provides a basis from which to measure participant progress, to identify the need for program enhancements, and to identify areas in which the program is effectively addressing participant needs.

Providing timely and integrated screening and assessments requires significant coordination among clinical and non-clinical staff, as well as the sharing of key pieces of information that can contribute to well-informed decision-making. Case staffings are used to share information regarding results of screening and assessment. Staff roles and responsibilities for screening and assessment should be clarified by drug court programs. In addition, training is needed to ensure that screening and assessment information is interpreted appropriately. For instance, screening personnel who have not been trained in substance abuse issues may exclude potential participants if no history of drug offenses exists, while in reality, some criminal defendants support their drug use by committing property offenses. For many drug court programs, designing a comprehensive screening and assessment system often takes a year or more. Once established, the system should be monitored and reevaluated periodically to ensure that it is appropriate for the target population, that staff is adequately trained, that protocols are followed, and that baseline data related to progress and outcome measures are adequately captured and analyzed.

Performance Benchmarks for Drug Court Screening and Assessment

Guidelines were developed by the National Association of Drug Court Professionals (NADCP) and the U.S. Department of Justice, Office of Justice Programs (*Defining Drug Courts: The Key Components*, 1997) that describe the best practices in developing and implementing drug court programs. Two of the "key components" contained in the guidelines refer to screening and assessment, including recommendations that "eligible participants are identified early and promptly placed in the drug court program," and that "drug courts provide access to a continuum" of services that include screening and assessment.

Performance benchmarks for each of the key components were developed to provide more specific guidance. Benchmarks that relate to screening and assessment include:[7]

- Eligibility screening is based on established written criteria. Criminal justice officials or others (e.g., pretrial services, probation, TASC)[8] are designated to screen cases and identify potential drug court participants.

- Eligible participants for drug court are promptly advised about program requirements and the relative merits of participating.

- Trained professionals screen drug court-eligible individuals for AOD[9] problems and suitability for treatment.

- Initial appearance before the drug court judge occurs immediately after arrest or apprehension, to ensure program participation.

- The court requires that eligible participants enroll in AOD services immediately.

- Individuals are initially screened and later periodically assessed by both court and treatment personnel to ensure that treatment services and individuals are suitably matched:

  - Ongoing assessment is necessary to monitor progress, to change the treatment plan as necessary, and to identify relapse cues.

  - If various levels of treatment are available, participants are matched to programs according to their specific needs. Guidelines for placement in various levels of treatment should be developed.

  - Screening for infectious diseases and health referrals occurs at an early stage.

The benchmarks set standards that drug courts should strive to meet. This guideline is intended to provide additional detail needed to support the achievement of these benchmarks.

---

[8]Treatment Accountability for Safer Communities (TASC) is a program model for assessing, referring to treatment, and providing case management services to substance abusers in the criminal justice system.

[9]AOD is an abbreviation for Alcohol and Other Drugs.

---

[7]The following material is quoted directly from the Key Components, referenced above, pp. 13, 16-17.

Drug Court Screening

Who Should Conduct Screening?

As this guideline indicated previously, several different drug court professionals may be involved in the clinical screening process. These individuals should be familiar with criminal justice processes, substance abuse treatment, admission criteria for the drug court program, the key components, and the requirements of the program. The screening interview is likely to be the potential participant's first contact with the drug court program, and it provides an important opportunity for staff to dispel myths about the program, to discuss ambivalence about recovery, to clarify potential treatment goals, and to mobilize optimism about involvement in the drug court program. Staff should be prepared to address a wide range of questions from potential program participants.

Although screening staff need not have extensive experience or training in assessment or diagnosis, they should receive training in substance abuse and treatment issues, basic interviewing skills, identification of mental health symptoms and warning signs for suicide, techniques for exploring motivation for treatment (e.g., motivational interviewing), and referral/triage to jail and community services. Knowledge of common "street" drugs, their use, and associated terminology is also important.

Training in substance abuse, interviewing, and basic counseling is often available through the Single State Agency that administers funding for drug and alcohol treatment. (A listing of Single State Alcohol and Drug Agency Directors is included as Appendix D.) In addition, local criminal justice agencies often provide training to their staff in substance abuse, drug testing, and interviewing techniques. Local universities also may be a resource for staff training.

Steps in Conducting Screening

Drug court screening should be completed at the earliest possible point after arrest, to expedite involvement in treatment and to capitalize on motivation for behavior change associated with the arrest. Program eligibility requirements should be written, clearly defined, and reviewed by all drug court staff. Once developed, eligibility criteria are sometimes translated into checklists for use by various screening staff. Drug courts that receive federal funding through the U.S. Department of Justice, Drug Courts Program Office, are also prohibited from admitting violent offenders. (Section 2201 of the Omnibus Crime Control and Safe Streets Act, 42 U.S.C. 3796ii). Eligibility criteria may also restrict admission of persons who have characteristics that may inhibit their successful involvement in a drug court program, such as infectious disease or active mental health symptoms.

Key aspects of the drug court program should be discussed at the time of the initial screening interview, including the duration of the program, the need for immediate detoxification services, the possibility of involvement in residential treatment, frequency of required treatment activities, specific hours that treatment services are offered, location of treatment facilities, drug testing, and the consequences of nonparticipation and unsuccessful termination. A written description of the services and requirements of the drug court program should also be provided. Discussion of program services, consequences of prior substance abuse and criminal justice involvement, and individual recovery goals provides an important opportunity to develop commitment to treatment. If the individual shows interest in the drug court program, a written consent should be signed. The consent form should include a description of information that will be shared, names of staff who will receive this information, and under what circumstances information will be shared.

If the screening interview occurs before formal acceptance into the drug court program, appropriate releases should be signed to enable screening staff to communicate with the court and other relevant individuals or agencies, to gather additional information, and to discuss the case. A properly executed and signed release of information must be completed for all drug court participants. A more detailed discussion of release of information is provided later in this monograph.

Several steps for screening participants in the drug court program are described as follows (see Belenko, 1996; Cooper, 1997; Peters et al., 1994), although the sequence may differ by program.

Justice System Screening

- Review new jail admissions or new arrest records for legal and statutory requirements to determine program eligibility. The prosecution and defense usually make initial legal screening decisions based on eligibility criteria developed by the drug court team. Areas commonly reviewed include:

  - Current charge(s),

  - Criminal history,

  - Circumstances of the current offense (e.g., defendant culpability, mandatory incarceration statutes, plea bargaining restrictions), and

  - Outstanding warrants, detainers, additional charges, or previous diversions that would disqualify the individual from participation in the drug court program.

This information is available from police and other criminal justice records. Other factors such as bail status of the individual, history of failure to appear in court, and history of incarceration may also influence this first-level eligibility determination. Potential participants who do not pass this first level of screening are not ordinarily reviewed further for the drug court program.

- Review by the defense attorney of complaint and discovery materials, the need for treatment, and the client's desire to seek treatment. The defense attorney will also describe the legal ramifications of participation in the drug court program, and will review waiver of rights to speedy trial.

- In most jurisdictions, the prosecutor and defense attorney sign off on the placement of a client in the program. In other jurisdictions, the judge or other team members may be involved.

Clinical Screening

- Interview of potential participant by pretrial services, probation, TASC, treatment staff, or other screening staff. Criminal justice records and other archival records may be reviewed before the interview. Screening instruments are often administered at the time of the interview. The interview should examine whether the individual has a substance abuse problem, if these problems can be meaningfully addressed within the drug court program, and if the individual is willing to comply with the requirements of the drug court program. Recommendations to the court should be presented in a neutral and unbiased way that balances a defendant's need for treatment with public safety and other goals of the criminal justice system.

Several issues that are often addressed during the interview include:

- The severity of the substance abuse problem, and whether treatment is warranted. The majority of drug court programs surveyed (Cooper, 1997) report that individuals who do not have an addictive disorder, or who only have a minimal substance abuse problem are not eligible for admission to the program.

- Whether the individual is a drug dealer or manufacturer. If the individual is either, and if individuals with these charges are eligible for the drug court, whether this would adversely affect involvement in treatment or would otherwise affect the treatment program or constitute too great a public safety risk.

- Willingness to participate in the drug court program, and agreement to comply with program requirements.

  Note: The drug court management team should discuss issues involving eligibility of persons who may be selling drugs. Some drug courts choose to include persons who are selling drugs as part of a drug-consuming lifestyle; other jurisdictions choose to include only persons charged with possession.

- Availability of services to meet the individual's needs for substance abuse treatment.

- Overriding issues that would prevent the individual's participation in the drug court program (e.g., pending charges that would require incarceration, mental illness or retardation that seriously impairs functioning) or other factors that cannot be addressed in available services.[10]

---

[10] Potential candidates with some of these attributes may be *good* candidates for admission to drug court programs that are structured for sentenced offenders, and that have access to comprehensive treatment and case management services. (See Peyton, E. and R. Gebelein, *TASC and Drug Courts: Natural Allies*, National TASC, Silver Spring, MD, 1995.)

Placement in Drug Court

- Final eligibility review by the judge and/or drug court staff.

- Referral of eligible participants to the next drug court session.

What Information Should Be Included in a Drug Court Screening?

Screening often includes a brief interview, the use of self-report instruments, and a review of archival records. When possible, it should also include recent results from drug tests. The type of screening information compiled by drug courts depends on the stage at which screening is conducted. Many programs use a short self-report instrument to document the frequency of drug and alcohol use over the past 30 days, and over a longer interval.

The following section describes several types of information that may be examined during screening for drug court programs, including core elements and other more specialized information.

Core Screening Elements

Background and Demographic Information

- Name, address, age, race/ethnicity, and gender

- Identifying numbers used by the court or the treatment provider

Criminal Justice Information

- Criminal history (prior felonies, violent offenses)

- Most recent offense of record

- Outstanding warrants, detainers, previous diversions, or other charges

Substance Use

- Signs of acute drug or alcohol intoxication

- Acute signs of withdrawal from drugs or alcohol

- Drug tolerance effects

- Results of recent drug testing

- Self-reported substance abuse

  • Age and pattern of first substance use

  • History of use

  • Current pattern of use (e.g., quantity, frequency, method of use)

  • "Drug(s) of choice" (including alcohol)

  • Motivation for using

- Negative consequences associated with substance use. For women, this may include changes in physical appearance.

- Prior involvement in treatment

- Family history of substance abuse (include family of origin as well as current family)

- Other observable signs and symptoms of substance abuse (e.g., needle marks/ injection sites, impaired motor skills)

  Note: Drug courts should develop clear policies regarding the use of alcohol by participants in drug court programs, as well as regarding the use of prescription drugs such as antidepressants and painkillers.

Mental Health

- Acute mental health symptoms (e.g., depression, hallucinations, delusions)

- Suicidal thoughts and behavior

- Other observable mental health symptoms

- Age at which mental health symptoms began

- Prior involvement in mental health treatment, and use of psychotropic medication

- Cognitive impairment

- Past or recent trauma such as sexual/ physical abuse

- Family history of mental illness

Other Indicators

- Motivation and readiness for substance abuse treatment

- Perceived level of substance abuse problems

- Infectious disease

- Social factors (e.g., primary responsibility for children, living with an abusive or substance-involved partner, sole economic provider responsibilities) that may present obstacles for treatment participation

Screening Issues for Women

Women present several unique issues that require additional consideration during screening and assessment. Many female offenders have a history of physical and sexual abuse, and have relationships characterized by unhealthy dependencies and poor communication skills (American Correctional Association, 1990; Lord, 1995). Mental health problems occur disproportionately among female offenders, particularly depression and post-traumatic stress disorder (Peters et al., 1997; Teplin et al., 1996). Many women also have responsibility for minor children (Bureau of Justice Statistics, 1994). Despite these unique needs, few jurisdictions offer specialized treatment services for women.

Key Points

- Many screening instruments were developed for males, and may not include questions that address issues relevant to women.

- Barriers to treatment participation should be identified, including responsibility for the care and support of minor children and other child custody issues.

- Circumstances related to housing and relationships should be examined to ensure that the woman is safe in her current living situation and that there are no pressures from significant others to continue drug or alcohol use. If the woman is in a situation where she is at risk for abuse, steps should be taken to develop a safety plan.

- Factors that led to prior relapses should be explored.

- Current and prior mental health diagnoses and treatment needs should be identified, and women should be asked if they are currently taking medication for anxiety, depression, etc.[11]

Screening for Mental Health Problems

Due to the high rates of mental health disorders among criminal justice populations, mental health symptoms and status should be routinely examined in drug court screenings.

Key Points

- Drug court programs should strive to be inclusive in admitting individuals with mental health disorders and other potentially disabling conditions (e.g., physical disabilities).

- Many individuals with mental health problems have successfully participated in drug court programs throughout the country.

- Drug courts should not restrict admission solely based on mental health symptoms or a history of mental health treatment, but should instead consider the degree to which mental health or other disorders lead to functional impairment that inhibits effective program participation.

- Key mental health indicators that may inhibit functioning in the drug court program include the following:

  - Paranoia, hallucinations, delusions, severe depression, or mania (i.e., hyperactivity and agitation) that occurs frequently, is obvious to others, is disruptive to group activities, or otherwise prevents constructive interaction with drug court staff or participants;

  - Lack of stabilization on psychotropic medication, or failure to follow medication regimes; and

  - Suicidal thoughts or other behavior.

- Each drug court should evaluate its capacity to work with participants with mental health problems. This evaluation should include examining existing program resources and other community mental health services, and identifying levels of functioning needed to participate effectively in those programs.

- Screening staff need to be trained to be knowledgeable in the identification of mental health symptoms, the nature and course of mental health disorders, commonly prescribed psychotropic medications, and referral procedures for mental health services.

---

[11]For additional information on substance abuse and women, see Covington, S., *Helping Women to Change in Correctional Settings*, San Francisco, CA: Jossey-Bass. In press.

Screening for Suicide

Screening for suicide risk should be a priority in all drug courts, because individuals who have recently been arrested and have substance use disorders have higher rates of suicidal behavior.

Note: Drug courts should have clear policies and procedures for handling participants who exhibit suicidal behavior. Substance abuse and mental health treatment programs, the Single State Agency, the State Mental Health agency, and local correctional and behavioral health agencies can provide guidance in this area.

Key Points

- Ongoing suicide screening should be provided for all potential drug court participants. While suicide screening is important for all drug court participants, it is particularly important for those with mental health disorders and those with a history of childhood abuse.

- All suicidal behavior (including threats and attempts) should be taken seriously and assessed promptly to determine the type of immediate intervention needed.

- Suicide screening is particularly important among participants who have severe depression or schizophrenia, or who are suffering from stimulant withdrawal.

- Screening should address the following areas:

  • Current mental health symptoms,

  • Current suicidal thoughts, and

  • Previous suicide attempts and their seriousness.

Useful Questions in Screening for Suicide

- How specific is the plan?

- What method will be used?

- When will it happen?

- How available are potential instruments (drugs, weapons)?

Screening for Motivation and Readiness for Treatment

Drug court screening and assessment should address an individual's motivation and readiness for treatment. Motivation may be affected by perceived sanctions and incentives, and may increase when continued substance abuse threatens current housing, involvement in mental health treatment, vocational rehabilitation, family (including loss of children), or marriage, or may lead to incarceration. Apparent lack of motivation should not, as a singular factor, be used to disqualify candidates from admission to the drug court program or to treatment, unless the candidate refuses to participate.

Research has shown that treatment outcomes for persons coerced or court-ordered to treatment are as good as or better than for participants in voluntary treatment (DeLeon, 1988; Hubbard et al., 1989; Leukefeld and Tims, 1988; Platt et al., 1988). Although some offenders may initially agree to participate in treatment to reduce negative consequences, motivation for treatment is expected to become internalized over time. Individuals often cycle through the following "stages of change" during the treatment and recovery process (Prochaska et al., 1992):

- Precontemplation (unawareness of problems),

- Contemplation (awareness of problems),

- Preparation (reached a decision point),

- Action (actively changing behaviors), and

- Maintenance (practices ongoing preventive behaviors).

14

Individuals in the earliest stages of change have little awareness of substance abuse (or other) problems, and no intentions of changing their behavior. Awareness of problems increases in later stages, as the individual begins to consider the goal of abstinence. Due to the chronic relapsing nature of substance abuse problems, movement through stages of change is not a linear process.

One function of the drug court program is to motivate participants toward recovery. By using sanctions and rewards, the judge can use the leverage of the criminal justice system to facilitate reductions and cessation of drug use. Although participants may enter drug court programs to avoid criminal penalties, the process of treatment can help to instill internal motivation needed for long-term change. While drug court participants will frequently return to previous stages of change before achieving sustained abstinence, drug court methods can reduce the likelihood that relapse will go unchecked and can encourage movement to more advanced stages of recovery.

Key Points

- Treatment is likely to be ineffective until individuals accept the need for treatment of substance abuse problems.

- Placement in different types of drug court services based on the participant's current motivation level is likely to enhance treatment compliance, retention, and outcomes.

  - Assessment of stages of change is useful in treatment planning, and in matching the individual to different types of treatment.

  - For individuals in early stages of change, placement in treatment that is too advanced, and that does not address a participant's ambivalence regarding behavior change, may lead to drop out from treatment.

- For individuals in later stages of change, placement in services that focus primarily on early recovery issues may also lead to drop out from treatment.

- Several instruments have recently been developed to examine motivation and readiness for treatment.

Useful Questions in Screening for Treatment Motivation and Readiness

- Do you have problems related to your alcohol or drug use? How serious do you think your alcohol or drug problems are?

- Do you want to make changes in your alcohol or drug use?

- Have you taken any steps to reduce your alcohol or drug use?

- How important is it for you to get treatment for your alcohol or drug problems?

Use of Self-Report Information

Most screening and assessment in drug courts is based on self-reported information. While self-reported information has generally been found to have good reliability and specificity, with criminal justice populations it is widely accepted that collateral information and chemical testing should supplement self-reported information. Self-reported information may be limited due to the following considerations:[12]

- Individuals in the criminal justice system may underreport mental health or substance abuse problems if they believe that accurate reporting may lead to involvement in highly structured, lengthy, or otherwise difficult treatment programs or criminal sanctions;

---

[12]Adapted from Peters and Bartoi (1997).

15

- Current and past use may be minimized because of denial and failure to perceive the relationship between substance use and related problems;

- Mental health disorders may interfere with the accuracy of responses;

- Cognitive impairments may impede screening and assessment;

- Effects of acute intoxication, withdrawal effects, or chronic substance abuse may limit the ability to provide accurate self-reported information; or

- A chronic history of substance abuse contributes to difficulties in remembering dates, onset, and effects of the disorder.

Strategies to Enhance the Accuracy of Self-Reported Information

- Supplement interview and test results with information from collaterals.

- Examine archival records to determine the onset, course, diagnoses, and responses to treatment of substance use and mental health disorders.

- Provide regular drug testing.

- Wait to use self-report instruments until it is determined that an individual is not in withdrawal or intoxicated.

- Provide repeated screening and assessment on a regular basis.

- Provide a supportive interview setting.

  - Self-reported information should be compiled in a non-judgmental manner, and in a relaxing setting when possible.

  - The interview should be prefaced with a discussion of the limits of confidentiality.

- Use motivational interviewing techniques, including:

  - Express empathy.

  - Develop discrepancy between stated goals and current behaviors (e.g., desire to keep a steady job vs. "binge" drug use).

  - Avoid arguing.

  - Roll with resistance by offering new ideas and finding new ways to encourage behavior change.

- Support self-efficacy, or self-confidence.

What Instruments Should Be Used in Drug Court Screening?

Drug courts should use standardized substance abuse screening instruments to enhance the consistency and validity of results. Approximately 75 percent of drug courts use standardized instruments for clinical screening and assessment, according to a recent nationwide survey (Cooper, 1997). The most commonly used instruments were the Addiction Severity Index (ASI), the Substance Abuse Subtle Screening Inventory (SASSI), the Michigan Alcoholism Screening Test (MAST), and the Offender Profile Index (OPI). Several of these instruments are discussed in more detail in the following sections.

Screening instruments should be administered concurrently with an individual interview, drug testing (if possible), and examination of collateral information. As described previously, drug court screening instruments should address the following key components: (1) symptoms of alcohol and drug abuse/dependence, (2) patterns of recent and current substance abuse, (3) signs and symptoms of major mental health disorders (e.g., depression, bipolar disorder, schizophrenia), (4) suicide risk, and (5) other motivational and health factors that may affect involvement in treatment. Use of objective "risk assessment" scales to examine public safety risk may also be administered at the time of screening.

Given the absence of current instruments that address each of these components, several independent instruments are often combined in screening. Examples of selected instruments in several different content areas are described in the section to follow.

Key Issues in Selecting Screening Instruments

Instruments used in screening for substance abuse treatment programs differ significantly in their coverage of substance abuse symptoms and mental health symptoms, validation for use in criminal justice and other settings, cost, scoring procedures, and training required for administration and scoring. Several key issues that should be addressed in selecting screening instruments include the following:

- **Reliability.** Reliability refers to the consistency of results obtained over time.

- **Validity.** In the area of screening, validity refers to the extent to which instruments can identify substance abuse problems effectively. The validity of screening and assessment instruments varies significantly. For example, many standardized substance abuse instruments do not adequately identify individuals with substance abuse problems, or are unable to identify individuals who do not have substance abuse problems.

- **Use in Criminal Justice Settings.** Few substance abuse instruments have been validated within criminal justice settings (Peters, 1992; Peters and Greenbaum, 1996).

- **Cost.** Several commercially available screening instruments have been heavily marketed to the substance abuse treatment

community in recent years. However, recent research (see section to follow) shows that several public domain instruments are among the most effective for use with criminal justice populations.

Substance Abuse Screening Instruments

Many screening instruments are currently in use in drug courts. While numerous instruments are available, few studies have examined the validity of different substance abuse screening instruments in criminal justice settings. In the most comprehensive study of this type (Peters and Greenbaum, 1996), three screening instruments were found to be the most effective in identifying prison inmates with substance dependence problems:

- ADS/ASI – Drug (a combined instrument, consisting of the Alcohol Dependence Scale and the Addiction Severity Index – Drug Use section; Skinner & Horn, 1984; McLellan et al., 1980)

- TCU Drug Dependence Screen (DDS; Simpson et al., 1997)

- Simple Screening Instrument (SSI; Center for Substance Abuse Treatment, 1994a)

These instruments outperformed several other substance abuse screens, including the Michigan Alcoholism Screening Test (MAST) — Short version; the ASI — Alcohol Use section, the Drug Abuse Screening Test (DAST-20); and the Substance Abuse Subtle Screening Inventory (SASSI-2) on key validity measures. The ADS/ASI, DDS, and SSI appear to hold considerable promise for use with participants in drug court programs. Copies of the DDS and SSI, along with instructions for use are included in Appendix A.[13] Information regarding availability and cost of additional instruments has been included in Appendix B.

---

[13] The ADS has not been included because it is a commercially available instrument.

In community settings,[14] several screening instruments have been found to have adequate validity for use with substance-abusing populations (McHugo et al., 1993; Peters and Greenbaum, 1996; Ross et al., 1990; Staley and El Guebaly, 1990). These include the Alcohol Dependence Scale (ADS), the Drug Abuse Screening Test (DAST; and DAST-20, a short version of the DAST), the Michigan Alcoholism Screening Test (MAST; and SMAST — a short version of the MAST), and the CAGE.

Mental Health Screening Instruments

Several brief mental health screens are available (e.g., BSI, RDS, SCL-90-R) that examine a broad range of mental health symptoms, while others focus on symptoms of a single disorder, such as depression (e.g., BDI). Information related to cost and availability is included in Appendix B. Several commonly used screening instruments that have been validated for use in detecting mental health symptoms[15] are described as follows:

- Beck Depression Inventory (BDI; Beck and Beamesderfer, 1974)

- Brief Symptom Inventory (BSI; Derogatis and Melisaratos, 1983)

- Referral Decision Scale (RDS; Teplin and Schwartz, 1989)

- Symptom Checklist 90 — Revised (SCL-90-R; Derogatis et al., 1974)

---

[14] Instruments developed for incarcerated offenders attempt to account for the interruption in drug use that occurs due to incarceration.

[15] See also, Peters and Bartoi (1997) for a more complete description of mental health screening instruments used in criminal justice settings, Allen and Columbus (1995), Center for Substance Abuse Treatment (1994b), and Rounsaville et al. (1996).

Motivational Screening Instruments

Several instruments are available that examine motivation and readiness for treatment. These instruments are designed primarily to identify individuals for whom admission to substance abuse treatment is inappropriate. Two of these instruments (SOCRATES, URICA) are based on the "stage of change" model. As described previously, information regarding motivation and readiness for treatment has been found to predict drop out from treatment and treatment outcome, and may be particularly useful in matching individuals to different types of treatment services provided by drug court programs. Instruments that examine motivation and readiness for treatment include the following:

- Circumstances, Motivation, Readiness, and Suitability Scale (CMRS; DeLeon and Jainchill, 1986)

- Stages of Change Readiness and Treatment Eagerness Scale (SOCRATES; Miller, 1994)

- University of Rhode Island Change Assessment Scale (URICA; McConnaughy, et al., 1983; DiClemente and Hughes, 1990)

What Screening Information Is Most Relevant to the Court?

The report prepared after completion of screening often contains the first set of descriptive information that the court will receive. As such, it gives the judge the opportunity to engage with participants in a meaningful way. In addition, information from the screening process will enable the judge and other members of the drug court team to decide whether the drug court program is appropriate for the participant, or whether another criminal justice intervention might be more appropriate. Screening

18

information that is most relevant to the court includes:

- Whether the defendant meets criminal justice criteria for admission, including current offense and criminal history;

- Whether the defendant is a good risk for community placement;

- Psychosocial history, including employment status, educational status, significant relationships, and living arrangements;

- Level of substance abuse involvement, and whether or not there is appropriate and available treatment to address the substance abuse problems;

- Willingness to comply with the requirements of the drug court program;

- Mental health symptoms that may prevent effective program participation; and

- History of failure to appear for court; prior probation and substance abuse treatment history.

Note: Additional information may be required or may be useful in certain jurisdictions. The drug court management team should develop policies and procedures regarding information to be presented to the court at the time of the initial drug court appearance, and regarding who will provide the information. In addition, the court's response to key issues identified in the screening process should be discussed. For instance, prior failure in treatment need not disqualify an individual from participation in the drug court program; multiple treatment episodes is sometimes necessary to accomplish sustained recovery.

Drug Court Assessment

Assessment explores many of the same issues as screening does, but in much more depth and with a particular emphasis on problem areas highlighted during screening. The purpose of assessment is not to determine eligibility but to develop a treatment plan and to decide the timing and application of specific services and programs. Assessment provides the basis for development of an individualized treatment plan or case management plan and for matching participants with different types of drug court services. Key elements of drug court assessment include substance abuse history, current patterns of use, mental health history and current symptoms, criminal justice history and status, other areas of psychosocial functioning, current skills deficits, and types of treatment and ancillary services needed. Standardized assessment instruments and methods should be used by drug court programs (National Institute of Corrections, 1991; Peters, 1992; U.S. Department of Justice, 1997).

When Should Drug Court Assessment Be Conducted?

Assessment is usually accomplished following completion of screening and following initial admission to the program. Sufficient time should be provided before initial assessment to ensure that an individual is detoxified and sober. Time will also show if any mental health symptoms are related to withdrawal from substance use (Weiss and Mirin, 1989). Although the initial assessment is often conducted in the first several weeks of the drug court program, assessment is an ongoing process, and must consider new issues that arise, and new information obtained during treatment. For example, prior physical and sexual abuse are often not reported until an individual is comfortable in revealing sensitive information to treatment counselors

and other treatment participants. Relapses that occur during treatment, changes in living arrangements and employment, and other new issues are often reviewed by a drug court treatment team, with modifications then made to the treatment plan to reflect new problem areas and related services provided to address these problems.

Note: Drug courts should develop a plan for managing participants or potential participants who have not achieved sobriety prior to admission. Many drug courts have access to residential or outpatient detoxification facilities; some defendants are detoxified in jail if they are unable to achieve sobriety in the community.

Who Should Conduct Assessment?

Over half of drug court assessments are conducted by a private treatment agency affiliated with the program, and 38 percent of drug courts use more than one agency to conduct assessments (Cooper, 1997). Drug court assessments should be conducted by professionals with experience and training in substance abuse treatment, diagnosis, and basic counseling techniques. These individuals should also have experience and training in criminal justice processes and in working with offenders. Assessments are typically conducted by certified substance abuse or addiction counselors, social workers, psychologists, and clinical nurse specialists. Licensed medical practitioners can provide assessment and diagnosis of health-related disorders, and can conduct routine physical examinations. For example, psychiatrists may provide consultation in examining individuals for mental health disorders and determining the need for psychotropic medication. States vary in their requirements regarding the qualifications of those who may conduct clinical assessments. The Single State Agency that administers federal funds for alcohol and drug treatment can provide information regarding these regulations. It is important for drug courts to use programs and counselors that meet local criteria for licensing and certification.

What Information Should Be Included in a Drug Court Assessment?

The following types of information should be examined in a drug court assessment, with particular emphasis given to those areas identified as problematic during screening:

- Criminal justice history and status

- Substance abuse history, current symptoms, and level of functioning

- Mental health history, current symptoms, and level of functioning

- History of interaction between mental health and substance use disorders

- Family history of substance use disorders (including birth complications and in utero substance exposure)

- Medical and health status

- Social/family relationships (including involvement in domestic violence and child abuse or neglect)

- Employment/vocational status

- Educational history and status

- Literacy, IQ, and developmental disabilities

- Treatment history and response to/compliance with treatment

- Prior experience with peer support groups

- Cognitive appraisal of treatment and recovery

  • Motivation and readiness for treatment

  • Self-efficacy in adopting lifestyle changes (e.g., maintaining abstinence, complying with medication)

  • Expectancies related to substance use (both positive and negative)

- Participant conceptualization of treatment needs

- Resources and limitations affecting the ability to participate in treatment (e.g., transportation problems, homelessness, child care needs)

- Interpersonal coping strategies, problem solving abilities, and communication skills

Areas for Detailed Assessment

Substance Abuse History and Status

- Substance abuse information should include the drug(s) of first preference; other secondary drugs; misuse of prescription drugs; age of onset; frequency, amount, and duration of current and past use; patterns of high and low usage; reasons for substance abuse; context of substance abuse, including methods of financing substance use; periods of abstinence and how they were attained; and information regarding prior relapses (e.g., antecedents, warning signs, and high risk situations).

- Assessment should examine the number and type of prior treatment experiences (e.g., whether treatment was voluntary or was the result of civil or criminal commitment), and treatment outcomes.

Mental Health History and Status

- Approximately 40-50 percent of substance-abusing offenders have a major mental illness (National GAINS Center, 1997; Teplin, 1994; Teplin et al., 1996).

- Mental health information should include current and past symptoms (e.g., suicidal behavior, depression, anxiety, psychosis, paranoia, stress, self-image, inattentiveness, impulsivity, hyperactivity), treatment history, use of psychotropic medications, and patterns of denial and manipulation.

- Mental health symptoms should be examined to determine whether the individual can function adequately in a drug court setting, the level of supportive services needed (e.g., mental health counseling, psychiatric consultation), and the need for a more thorough mental health assessment by a psychologist or psychiatrist.

Family and Social Relationships

- Assessment should examine social interactions and lifestyle, effects of peer pressure to use drugs and alcohol, and available peer and family support for involvement in treatment. This area of assessment is particularly important with juveniles.

- The history of abuse and neglect within the family should be examined, in addition to the family history of substance abuse, mental illness, and criminal justice involvement.

- The stability of the home and social environment should also be assessed, including violence in the home and effects of the home and other relevant social environments (e.g., work, school) on abstinence from substance use.

- The history of marital and other significant relationships, important life events, and childhood history should be examined.

Medical/Health Care History and Status

- Key areas to examine include history of injury and trauma, chronic disease, physical disabilities, substance toxicity and withdrawal, impaired cognition, neurological symptoms, and prior use of medication.

- If a history of Attention Deficit or Hyperactivity Disorders (AD/HD) is suspected, the assessment should examine attention and concentration difficulties, hyperactivity and impulsivity, and the developmental history of childhood AD/HD symptoms.

Criminal Justice History and Status

- The complete criminal history should be reviewed. The pattern of prior criminal offenses may reveal important information regarding the effect of substance abuse on criminal behavior, the need for case management services, and potential relapse prevention strategies (e.g., avoidance of specific high-risk situations that may elicit a return to criminal behavior and substance abuse). The self-reported history provided during assessment should be corroborated through inspection of official criminal justice history records.

Key assessment information related to criminal history includes the following:

- Prior arrests (including age at first arrest, type of arrest)

- Juvenile justice history

- Involvement with the civil justice system, including domestic violence, child abuse or neglect, custody issues, etc.

- Alcohol and drug-related offenses (e.g., DUI/DWI, drug possession or sales, reckless driving)

- Level of intoxication at the time of previous offenses (either reported or unreported offenses)

- Felony convictions

- Number of prior jail and prison admissions, duration of incarceration

- Disciplinary incidents in jail and prison

- Use of isolation management in jail and prison

- Probation or parole violations

■ The current criminal justice status should also be examined. This information will help in coordinating treatment and management issues with courts and community supervision staff.

Key assessment information related to current criminal status includes the following:

- Court orders requiring assessment and involvement in treatment, including the length of involvement in treatment (if specified)

- Duration of criminal justice supervision (e.g., pretrial release, probation, parole)

- Supervision arrangements (e.g., supervising probation/parole officer, frequency of court or supervision appointments, reporting requirements)

- Consequences for noncompliance with treatment guidelines

What Instruments Are Available for Assessment of Participants in Drug Court Programs?

Few comprehensive instruments have been validated for use in assessing individuals with substance abuse disorders, and no instrument is perfect. Moreover, few studies have attempted to validate the use of assessment instruments in criminal justice settings. A comprehensive approach should be developed to assess participants in drug court programs. This approach should include review of substance use and other disorders, examination of criminal justice history and status, and drug screens. One example of a comprehensive assessment approach is the Addiction Severity Index (ASI),[16] which is one of the few available instruments that measures several different functional aspects of psychosocial functioning related to substance abuse. The ASI provides a concise review of the history of substance abuse and recent use. The ASI is described in more detail in the following section. Several previously described screening instruments are often used as part of an assessment battery (e.g., to examine diagnostic symptoms of alcohol or drug abuse and dependency).

Addiction Severity Index (ASI)

The ASI (McLellan et al., 1980; McLellan et al., 1992) is currently the most widely used substance abuse instrument, and is used for screening, assessment, and treatment planning. The ASI is a "public domain" instrument developed through the National Institute on Drug Abuse (NIDA). The instrument provides a structured interview format to examine seven areas of functioning that are commonly affected by substance abuse, including drug/alcohol use, family/social relationships, employment/support status, and mental health status. Many agencies, including those in criminal justice settings, have modified the ASI for use as a screening instrument for substance abuse. Two independent sections of the ASI examining drug and alcohol use are frequently used as screening instruments.

---

[16] Appendix C includes a copy of the ASI instrument.

Key Features and Considerations

- The ASI has been found to be reliable and valid for use with a range of substance-abusing populations, including offenders, and is highly correlated with objective indicators of addiction severity (McLellan et al., 1985).

- Severity ratings are provided in each functional area assessed, which may be useful for clinical and research purposes.

- Staff training is needed to administer and score the ASI. Administration of the entire ASI requires 45-75 minutes.

- Although developed for use in an interview, a self-report version of the ASI (SA-ASI) has recently been developed. The psychometric properties of this self-report instrument have not yet been established.

What Assessment Information Is Most Relevant to the Court?

Assessment information can provide important guidance to the court regarding a participant's adaptation to treatment; strengths and weaknesses; supervision, management, and treatment strategies; and potential pitfalls to avoid during involvement in the program. This unique information is often pivotal in cementing the relationship between the drug court participant and the judge. Assessment information also helps identify key areas to monitor and review during drug court status hearings, and allows the judge to develop individualized sanctions and incentives.

The following assessment information is particularly useful to the court:

- Current placement and status or adjustment in treatment

- Treatments attempted, and the outcomes of these interventions

- Whether the participant lives in a drug-free and stable residence

- Whether significant others (e.g., spouses, coworkers, girlfriends/boyfriends, family members) are active substance abusers; whether significant others support recovery goals

- High-risk situations for substance abuse relapse

- Personal recovery goals

- Employment status and skills

- Mental health problems

- Medical problems

- History of violence or abuse (either as perpetrator or victim)

- Additional services that will be required by the participant

- Obstacles to participant progress

- Issues that may affect the participant's ability to remain or succeed in the program

Obtaining Release of Confidential Information

Federal Confidentiality Regulations (42 CFR Part 2) prohibit the release of information about participants in substance abuse treatment without a written consent from the individual (or the parent, if the participant is a minor, in areas in which treatment is contingent on parental consent). Confidentiality laws are fairly restrictive, and are designed to protect the privacy rights of participants in substance abuse treatment.[17] Violations of these regulations can result in substantial fines. All drug courts should become familiar with federal and state confidentiality regulations, and should develop procedures to ensure that cooperating agencies comply with these regulations.

Confidentiality regulations are generally not as strict for treatment participants who are supervised by the criminal justice system, and do not prohibit the exchange of information between affiliated drug court agencies, or with other criminal justice or community agencies. Once consent for release of information is provided within criminal justice settings, it generally cannot be rescinded until the participant graduates or leaves the program. Individuals who receive confidential information may disclose and use it only to carry out their official duties with respect to the release.

---

[17] For additional information, see Center for Substance Abuse Treatment (1994c), *Confidentiality of patient records for alcohol and other drug treatment*. Technical Assistance Publication (TAP) Series, #13, and Center for Substance Abuse Treatment (1994d), *Combining substance abuse treatment with intermediate sanctions for adults in the criminal justice system*. Treatment Improvement Protocol (TIP) Series, #12.

In general, release of information forms completed for participants in drug court programs should describe the following:

- The name of the participant in the drug court program

- The name or general designation of the individual who is permitted to disclose information

- Criminal justice staff who may receive the information in connection with their duty to monitor the participant's progress

- The purpose of the disclosure

- The type of information to be released

- The period during which the release remains in effect (e.g., anticipated length of participation in drug court treatment, anticipated duration of criminal justice supervision)

- The signature of the drug court participant and/or of the parent, as needed

- The date on which the release form was signed.

Summary

The number of drug courts implemented throughout the country has increased dramatically in the past five years. Drug courts offer significant new opportunities to effectively manage, supervise, and treat individuals with substance abuse problems, and to discourage their return to the criminal justice system, by blending criminal justice interventions with effective treatment methods and programming.

Screening and assessment activities are important to drug courts in identifying participants who meet eligibility criteria, and in selecting individuals who are likely to benefit from drug court intervention. Screening refers to the relatively brief examination of program eligibility criteria (both criminal justice and clinical), while assessment involves a more detailed review of psychosocial problems and treatment needs. Screening and assessment activities form the basis of the ongoing, individualized dialogue between the participant and the drug court team that characterizes drug courts.

Failure to provide appropriate screening and assessment can lead to misidentification of problems, ineffective treatment planning, and placement in services that are inconsistent with the needs of participants. An effective screening and assessment system can give drug courts the data they need to augment services, strengthen program weaknesses, and identify and build on program strengths.

Drug court programs should develop written eligibility criteria to guide the screening process. Policies and procedures should be developed that describe roles and responsibilities of staff involved in screening and assessment, information sharing, and methods to safeguard participant confidentiality. Eligibility criteria should be designed to permit participation of individuals with mental health and other potentially disabling disorders. Criteria related to these disorders should focus on functional impairment that would inhibit meaningful participation in the drug court.

Screening should be conducted as soon after arrest as possible, to expedite involvement in the drug court program. Both screening and assessment should be based on multiple sources of information, including interview, self-report instruments, and review of records. Screening and assessment for participants in the drug court program should examine various types of information related to substance abuse and mental health disorders and criminal justice involvement to form a comprehensive and integrated description of each participant's supervision and treatment needs.

Several instruments are available for both screening and assessment. Use of standardized screening and assessment instruments by drug courts will enhance the consistency and accuracy of results. Several substance abuse screening instruments have been validated recently for use in criminal justice settings, and appear to hold considerable promise for use in drug courts. Other instruments have been validated for use in examining mental health disorders.

Staff training is needed in the areas of interviewing strategies, identification of key indicators and problem areas, use of instruments, and referral to services, as well as other areas. Training is often available through Single State Agencies, local universities, providers of treatment programs for substance abuse or provider associations, and criminal justice agencies.

References

American Correctional Association (1990). *The female offender: What does the future hold?* Washington, DC: St. Mary's Press.

American Society of Addiction Medicine (1996). *Patient placement criteria for the treatment of substance-related disorders: Second Edition.* Chevy Chase, MD.

Beck, A.T., and Beamesderfer, A. (1974). Assessment of depression: The depression inventory. In P. Pichot (Ed.) *Modern Problems in Pharmacopsychiatry* (pp. 151-169). Basel, Switzerland: Karger.

Belenko, S. (1996). *Comparative models of treatment delivery in drug courts.* New York: The Sentencing Project.

Bureau of Justice Statistics. *Special Report: Women in Prison.* Washington, DC: U.S. Department of Justice, 1994.

Center for Substance Abuse Treatment (1994). *Confidentiality of patient records for alcohol and other drug treatment.* Technical Assistance Publication (TAP) Series, #13. Rockville, MD.

Center for Substance Abuse Treatment (1994b). *Screening and assessment for alcohol and other drug abuse among adults in the criminal justice system.* Treatment Improvement Protocol (TIP) Series, #7. Rockville, MD.

Center for Substance Abuse Treatment (1994c). *Simple screening instruments for outreach for alcohol and other drug abuse and infectious diseases.* Treatment Improvement Protocol (TIP) Series, #11. Rockville, MD: U.S. Department of Health and Human Services.

Center for Substance Abuse Treatment (1994d). *Combining substance abuse treatment with intermediate sanctions for adults in the criminal justice system.* Treatment Improvement Protocol (TIP) Series, #12. Rockville, MD.

Cooper, C.S. (1997). 1997 Drug Court Survey. Washington, DC: American University.

Covington, S. (1998). *Helping women to change in correctional settings.* San Francisco, CA: Jossey-Bass. In press.

DeLeon G. (1988). "Legal Pressure in Therapeutic Communities," Journal of Drug Issues, 18: 625-640.

DeLeon, G., and Jainchill, N. (1986). Circumstance, Motivation, Readiness and Suitability as correlates of treatment tenure. *Journal of Psychoactive Drugs, 18* (3), 203-208.

Derogatis, L., Lipman, R., and Rickels, K. (1974). The Hopkins Symptom Checklist (HSCL): A self-report symptom inventory. *Behavioral Science, 19*, 1-16.

Derogatis, L.R., and Melisaratos, N. (1983). The Brief Symptom Inventory: An introductory report. *Psychological Medicine, 13*, 595-605.

DiClemente, C.C., and Hughes, S.O. (1990). Stages of change profiles in outpatient alcoholism treatment. *Journal of Substance Abuse, 2*, 217-235.

Drake, R.E., Alterman, A.I., & Rosenberg, S.R. (1993). Detection of substance use disorders in severely mentally ill patients. *Community Mental Health, 29* (2), 175-192.

Hubbard, R.L., M.E. Marsden, J.V. Rachel, H.J. Harwood, E.R. Cavanaugh, and H.M. Ginzburg (1989). Drug Abuse Treatment: A National Study of Effectiveness. Chapel Hill, NC: University of North Carolina Press.

Kivlahan, D.R., Sher, K.M., and Donovan, D.M. (1989). The Alcohol Dependence Scale: A validation study among inpatient alcoholics. *Journal of Studies on Alcohol, 50,* 170-175.

Kofoed, L., Dania, J., Walsh, T., and Atkinson, R.M. (1986). Outpatient treatment of patients with substance abuse and coexisting psychiatric disorders. *American Journal of Psychiatry, 143,* 867-872.

Leukefeld, C.G., and F.M. Tims [eds.] (1988). Compulsory Treatment of Drug Abuse: Research and Clinical Practice. Rockville, MD: National Institute on Drug Abuse.

Lord, E. A prison superintendent's perspective on women in prison. *The Prison Journal* 75(2):257-269, 1995.

McConnaughy, E.A., Prochaska, J., and Velicer, W.F. (1983). Stages of change in psychotherapy: Measurement and sample profiles. *Psychotherapy: Theory, Research, and Practice, 20,* 368-375.

McHugo, G., Paskus, T.S., and Drake, R.E. (1993). Detection of alcoholism in schizophrenia using the MAST. *Alcoholism: Clinical and Experimental Research, 17* (1), 187-191.

McLellan, A.T., Kushner, H., Metzger, D., Peters, R.H., Smith, I., Grissom, G., Pettinati, H., & Argeriou, M. (1992). The Fifth Edition of the Addiction Severity Index. Journal of Substance Abuse Treatment, 9, 199-213.

McLellan, A.T., Luborsky, L., Cacciola, J., Griffith, J., Evans, F., Barr, H.L., & O'Brien, C.P. (1985). New data from the Addiction Severity Index: Reliability and validity in three centers. Journal of Nervous and Mental Disease, 173, 412-423.

McLellan, A.T., Luborsky, L., Woody, G.E., and O'Brien, C.P. (1980). An improved diagnostic evaluation instrument for substance abuse patients: The Addiction Severity Index. *Journal of Mental and Nervous Disease, 168* (1), 26-33.

Miller, W.R. (1994). *SOCRATES: The Stages of Change Readiness and Treatment Eagerness Scale.* Albuquerque: University of New Mexico, Department of Psychology.

National GAINS Center (1997). The prevalence of co-occurring mental and substance disorders in the criminal justice system. *Just the Facts* series. Delmar, NY.

National Institute of Corrections (1991). *Intervening with Substance-Abusing Offenders: A Framework for Action.* Washington, DC: U.S. Department of Justice.

Office of Justice Programs (1997). *Defining drug courts: The key components.* Drug Courts Program Office, in collaboration with the National Association of Drug Court Professionals, Drug Court Standards Committee. Washington, DC: U.S. Department of Justice.

Peters, R.H. (1992). Referral and screening for substance abuse treatment in jails. *Journal of Mental Health Administration, 19* (1), 53-75.

Peters, R.H., and Bartoi, M.G. (1997). *Screening and assessment of co-occurring disorders in the justice system.* Delmar, NY: The National GAINS Center.

Peters, R.H., and Greenbaum, P.E. (1996). *Texas Department of Criminal Justice/Center for Substance Abuse Treatment Prison Substance Abuse Screening Project.* Milford, MA: Civigenics, Inc.

Peters, R.H., Pennington, B., Wells, J.D., Rosenthal, L., and Meeks, J. (1994). *Treatment-based drug courts: Gearing up against substance abuse.* Alexandria, VA: State Justice Institute.

Peters, R.H., Strozier, A.L., Murrin, M.R. and Kearns, W.D. (1997). Treatment of Substance-Abusing Jail Inmates: Examination of gender differences. *Journal of Substance Abuse Treatment, 14* (4), 339-349.

Peyton, E. and R. Gebelein, *TASC and Drug Courts: Natural Allies*, National TASC, Silver Spring, MD, 1995.

Prochaska, J.O., DiClemente, C.C., and Norcross, J.C. (1992). In search of how people change: Applications to addictive behaviors. *American Psychologist, 47*, 1102-1114.

Ross, H.E., Gavin, D.R., and Skinner, H.A. (1990). Diagnostic validity of the MAST and the Alcohol Dependence Scale in the assessment of DSM-III alcohol disorders. *Journal of Studies on Alcohol, 51* (6), 506-513.

Rounsaville, B.J., Tims, F.M., Horton, A.M., and Sowder, B.J. (Eds.). (1996). *Diagnostic source book on drug abuse research and treatment.* Rockville, MD: National Institute on Drug Abuse.

Skinner, H.A., and Horn, J.L. (1984). *Alcohol Dependence Scale: User's Guide.* Toronto: Addiction Research Foundation.

Simpson, D.D., Knight, K., and Broome, K.M. (1997). *TCU/CJ Forms Manual: Drug Dependence Screen and Initial Assessment.*

Fort Worth: Institute of Behavioral Research, Texas Christian University.

Staley, D., and El Guebaly, N. (1990). Psychometric properties of the Drug Abuse Screening Test in a psychiatric patient population. *Addictive Behaviors, 15*, 257-264.

Teplin, L.A. (1994). Psychiatric and substance abuse disorders among male urban jail detainees. *American Journal of Public Health, 84* (2), 290-293.

Teplin, L.A., Abram, K.M., & McClelland, G.M. (1996). Prevalence of psychiatric disorders among incarcerated women, I: Pre-trial jail detainees. *Archives of General Psychiatry, 53*, 505-512.

Teplin, L.A., and Schwartz, J. (1989). Screening for severe mental disorder in jails. *Law and Human Behavior, 13* (1), 1-18.

U.S. General Accounting Office (1995). *Drug courts: Information on a new approach to address drug-related crime.* Briefing report to the Committee on the Judiciary, U.S. Senate, and the Committee on the Judiciary, House of Representatives, #GAO/GGD-95-159BR. Washington, DC.

Weiss, R.D., & Mirin, S.M., (1989). The dual diagnosis alcoholic: Evaluation and treatment. Psychiatric Annals, 19 (5), 261-265.

Other Related Resource Materials

Allen, J.P., and Columbus, M. (Eds). (1995). *Assessing alcohol problems: A guide for clinicians and researchers.* Bethesda, MD: National Institute on Alcohol Abuse and Alcoholism.

American Psychiatric Association (1996). *Practice guidelines.* Washington, DC.

Center for Substance Abuse Treatment (1996). *Treatment drug courts: Integrating Substance Abuse Treatment With Legal Case Processing.* Treatment Improvement Protocol Series, #23. Rockville, MD.

Center for Substance Abuse Treatment (1995). *Combining alcohol and other drug abuse treatment with diversion for juveniles in the justice system.* Treatment Improvement Protocol Series, #21. Rockville, MD.

Center for Substance Abuse Treatment (1995). *Planning for alcohol and other drug abuse treatment for adults in the criminal justice system.* Treatment Improvement Protocol Series, #17. Rockville, MD.

Center for Substance Abuse Treatment (1995). *The role and current status of patient placement criteria in the treatment of substance use disorders.* Treatment Improvement Protocol Series, #13. Rockville, MD.

Center for Substance Abuse Treatment (1994). *Assessment and treatment of patients with coexisting mental illness and alcohol and other drug abuse.* Treatment Improvement Protocol Series, #9. Rockville, MD.

Center for Substance Abuse Treatment (1994). *Screening and Assessment for Alcohol and Other Drug Abuse Among Adults in the Criminal Justice System.* Treatment Improvement Protocol Series, #7. Rockville, MD.

Center for Substance Abuse Treatment (1993). *Screening for infectious diseases among substance abusers.* Treatment Improvement Protocol Series, #6. Rockville, MD.

Center for Substance Abuse Treatment (1993). *Screening and assessment of alcohol- and other drug-abusing adolescents.* Treatment Improvement Protocol Series, #3. Rockville, MD.

Donovan, D.M., and Marlatt, G.A. (1988). *Assessment of addictive behaviors.* New York: The Guilford Press.

Miller, W. R., & Rollnick, S. (1991). *Motivational interviewing: Preparing people to change addictive behavior.* New York: The Guilford Press.

National Institute on Alcohol Abuse and Alcoholism (1995). *Assessing alcohol problems: A guide for clinicians and researchers.* NIAAA Treatment Handbook series #4. Bethesda, MD.

National Institute on Drug Abuse (1994). *Mental health assessment and diagnosis of substance abusers.* Clinical Report Series. U.S. Department of Health and Human Services, NIDA Office of Science Policy, Education and Legislation. Rockville, MD.

Appendix A:

Selected Instruments

Appendix B:

Availability and Cost of Screening Instruments

Mental Health, Substance Abuse, and Motivational Screening Instruments

Mental Health Screening Instruments
Health Screening Instruments

Instrument	Time to Administer	Cost	Source
Beck Depression Inventory (BDI)	5 mins., 21 items	Basic kit is $45	Psychological Corporation (800) 211-8378
Brief Symptom Inventory (BSI)	10 mins., 53 items	Basic kit is $87	NCS Assessments (800) 627-7271
Referral Decision Scale (RDS)	5 mins., 14 items	No cost	Published in Law and Human Behavior, 1989; (13) 1, 1-18.
Symptom Checklist 90-Revised (SCL-90R)	15 mins., 90 items	Basic kit is $87	NCS Assessments (800) 627-7271

Substance Abuse Screening Instruments
Abuse Screening Instruments

Instrument	Time to Administer	Cost	Source
Alcohol Dependence Scale (ADS)	5 mins., 25 items	Basic kit is $15	Marketing Services Addiction Research Foundation 33 Russell Street Toronto, Ontario M5S 2S1 (416) 545-6000
Addiction Severity Index (ASI) – Drug Use section	10-15 mins., 24 items	No cost	DeltaMetrics/TRI (800) 238-2433, or QuickStart Systems (214) 342-9020. Also published in various TIP monographs by the Center for Substance Abuse Treatment
Drug Dependence Screen (DDS)	5 mins., 15 items	No cost	TCU/CJ Forms Manual Texas Christian University Institute of Behavioral Research (817) 921-7226 *Instrument can be downloaded at: www.ibr.tcu.edu*

Instrument	Time to Administer	Cost	Source
Simple Screening Instrument (SSI)	5 mins., 16 items	No cost	Published in Center for Substance Abuse Treatment TIP #11. Order TIP through NCADI Clearinghouse at (800) 729-6689

Motivational Screening Instruments
Screening Instruments

Instrument	Time to Administer	Cost	Source
Circumstances, Motivation, and Readiness Scales (CMRS)	10 min., 42 items	No cost	Center for Therapeutic Community Research, National Development and Research Institutes (212) 966-8700 11 Beach Street New York, NY 10014-2114
Stages of Change, Readiness, and Treatment Eagerness Scale (SOCRATES)	5 mins., 19 items	No cost	Scott Tonigan, Ph.D. University of New Mexico 2350 Alamo SW Albuquerque, NM 87131 (505) 768-0214
University of Rhode Island Change Assessment Scale (URICA)	15 mins., 32 items	No cost	University of Rhode Island Cancer Prevention Research Center Kingston, RI 02881 (401) 874-2830 *Instrument can be downloaded at: www.uri.edu/research/cprc/measures.htm. Instrument is published in McConnaughy, DiClemente, Prochaska, and Velicer (1989), Psychotherapy, 26, 494-503*

Appendix C:

Addiction Severity Index

Appendix D:

Single State Alcohol and Drug Agency Directors

SINGLE STATE ALCOHOL AND
DRUG AGENCY DIRECTORS

ALABAMA

Mr. O'Neill Pollingue
Director
Division of Substance Abuse Services
Alabama Department of Mental Health
 and Mental Retardation
RSA Union Building
100 N. Union Street
Montgomery, Alabama 36130-1410
TEL (334) 242-3953
FAX (334) 242-0759

ALASKA

Mr. Loren A. Jones
Director
Division of Alcoholism and Drug Abuse
Alaska Department of Health and
 Social Services
240 Main Street
Suite 701
Juneau, Alaska 99811

Mailing Address

P.O. Box 110607
Juneau, Alaska 99811-0607
TEL (907) 465-2071
FAX (907) 465-2185

ARIZONA

Ms. Christy Dye
Acting Program Manager
Office of Substance Abuse
Division of Behavioral Health Services
Arizona Department of Health Services
2122 East Highland
Phoenix, Arizona 85016
TEL (602) 381-8999
FAX (602) 553-9143

ARKANSAS

Mr. Joe M. Hill
Director
Arkansas Bureau of Alcohol
 and Drug Abuse Prevention
5800 West 10th Street, Suite 907
Little Rock, Arkansas 72204
TEL (501) 280-4500
FAX (501) 280-4519

CALIFORNIA

Andrew M. Mecca, Ph.D.
Director
Department of Alcohol and Drug Programs
California Health and Welfare Agency
1700 K Street, Fifth Floor
Executive Office
Sacramento, California 95814-4037
TEL (916) 445-1943
FAX (916) 323-5873

COLORADO

Ms. Janet Wood
Director
Alcohol and Drug Abuse Division
Colorado Department of Human Services
4300 Cherry Creek Drive, South
Denver, Colorado 80222-1530
TEL (303) 692-2930
FAX (303) 753-9775

CONNECTICUT

Thomas A. Kirk, Jr., Ph.D.
Deputy Commissioner
Department of Mental Health
  and Addiction Services
410 Capitol Avenue, MS 14COM
P.O. Box 341431
Hartford, Connecticut 06134
TEL (860) 418-6958

DELAWARE

Ms. Renata Henry
Director
Delaware Health and Social Services Division
  of Alcoholism, Drug Abuse, and
  Mental Health
1901 North DuPont Highway
New Castle, Delaware 19720
TEL (302) 577-4461

FLORIDA

Kenneth A. DeCerchio, MSW, CAP
Assistant Secretary
Florida Department of Children and Families
Substance Abuse Program Office
1317 Winewood Boulevard
Building 3, Room 101Y
Tallahassee, Florida 32399-0700
TEL (904) 487-2920
FAX (904) 487-2239

GEORGIA

Elizabeth M. Howell, M.D.
Substance Abuse Program Chief
Division of Mental Health, Mental Retardation,
  and Substance Abuse
GA Department of Human Resources
Two Peachtree Street, NW, Fourth Floor
Atlanta, Georgia 30303-3171
TEL (404) 657-6419
FAX (404) 657-5681/6424

HAWAII

Ms. Elaine Wilson
Chief
Alcohol and Drug Abuse Division
Hawaii Department of Health
1270 Queen Emma Street, Suite 305
Honolulu, Hawaii 96813
TEL (808) 586-3962

IDAHO

Patricia Getty, M.Ed.
State Director
Bureau of Mental Health and Substance Abuse
Division of Family and Community Services
Idaho Department of Health and Welfare
450 West State Street
Boise, Idaho 83720

Mailing Address

PO Box 83720, Fifth Floor
Boise, Idaho 83720-0036
TEL (208) 334-5935
FAX (208) 334-6699

ILLINOIS

Mr. Nick Gantes
Acting Director
Illinois Department of Alcoholism
  and Substance Abuse
James R. Thompson Center
100 West Randolph Street, Suite 5-600
Chicago, Illinois 60601
TEL (312) 814-2291/3840
FAX (312) 814-2419

INDIANA

Patrick Sullivan, Ph.D.
Director
Bureau of Addiction Services
Division of Mental Health
Indiana Family and Social Services
  Administration
402 West Washington Street, Room W-353
Indianapolis, Indiana 46204-2739
TEL (317) 232-7816
FAX (317) 232-3472

IOWA

Ms. Janet Zwick
Director
Division of Substance Abuse
  and Health Promotion
Iowa Department of Public Health
321 East 12th Street
Lucas State Office Building, Third Floor
Des Moines, Iowa 50319-0075
TEL (515) 281-4417
FAX (515) 281-4535

KANSAS

Mr. Andrew O'Donovan
Commissioner
Alcohol and Drug Abuse Services
Department of Social and
  Rehabilitation Services
300 SW Oakley, Biddle Building, Second Floor
Topeka, Kansas 66606-1861
TEL (913) 296-3925
FAX (913) 296-0494

KENTUCKY

Mr. Michael Townsend
Director
Division of Substance Abuse
Kentucky Department of Mental Health
  and Mental Retardation Services
275 East Main Street
Frankfort, Kentucky 40621
TEL (502) 564-2880
FAX (502) 564-3844

LOUISIANA

Mr. Alton Hadley
Assistant Secretary
Office of Alcohol and Drug Abuse
Louisiana Department of Health and Hospitals
1201 Capitol Access Road, Fourth Floor

Mailing Address

P.O. Box 2790, BIN #18
Baton Rouge, Louisiana 70821-3868
TEL (504) 342-6717
FAX (504) 342-3931

MAINE

Ms. Lynn Duby
Director
Maine Office of Substance Abuse
Augusta Mental Health Complex
Marquardt Building, Third Floor
159 State House Station
Augusta, Maine 04333-0159
TEL (207) 287-2595/6330
FAX (207) 287-4334

MARYLAND

Mr. Thomas Davis
Director
Alcohol and Drug Abuse Administration
Maryland Department of Health and
  Mental Hygiene
201 West Preston Street, Fourth Floor
Baltimore, Maryland 21201-2399
TEL (410) 225-6925
FAX (410) 333-7206

MASSACHUSETTS

Ms. Mayra Rodriquez-Howard
Director
Bureau of Substance Abuse Services
Massachusetts Department of Public Health
250 Washington Street
Boston, Massachusetts 02108
TEL (617) 624-5111
FAX (617) 624-5185

MICHIGAN

Ms. Karen Schrock
Chief
Center for Substance Abuse Services
Michigan Department of Community Health
3423 N. Martin L. King, Jr. Boulevard
P.O. Box 30195
Lansing, Michigan 48909
TEL (517) 335-8808
FAX (517) 335-8837

MINNESOTA

Cynthia Turnure, Ph.D.
Director
Chemical Dependency Program Division
Minnesota Department of Human Services
444 Lafayette Road North
St. Paul, Minnesota 55155-3823
TEL (612) 296-4610
FAX (612) 296-6244/297-1862

MISSISSIPPI

Mr. Herbert Loving
Director
Division of Alcohol and Drug Abuse
Mississippi Department of Mental Health
1101 Robert E. Lee State Building
239 North Lamar Street, Eleventh Floor
Jackson, Mississippi 39201
TEL (601) 359-1288
FAX (601) 359-6295

MISSOURI

Mr. Michael Couty
Director
Division of Alcohol and Drug Abuse
Missouri Department of Mental Health
1706 East Elm Street
P.O. Box 687
Jefferson City, Missouri 65102-0687
TEL (573) 751-4942
FAX (573) 751-7814

MONTANA

Mr. Dan Anderson
Administrator
Addictive and Mental Disorders Division
1400 Broadway
Room C-118
Mailing Address
P.O. Box 202951
Helena, Montana 59620-2951
TEL (406) 444-2827
FAX (406) 444-4920

NEBRASKA

Mr. Malcolm Heard
Director
Division of Alcoholism, Drug Abuse,
  and Addiction Services
Nebraska Department of Health
  and Human Services Systems
Folsom and Prospector Place

Mailing Address

P.O. Box 94728
Lincoln, Nebraska 68509-4728
TEL (402) 471-2851, ext. 5583
FAX (402) 479-5162

NEVADA

Ms. Marilynn Morrical
Chief
Bureau of Alcohol and Drug Abuse
Rehabilitation Division
Department of Employment, Training,
  and Rehabilitation
505 East King Street, Room 500
Carson City, Nevada 89710
TEL (702) 687-4790
FAX (702) 687-6239

NEW HAMPSHIRE

Ms. Denise Devlin
Director
Office of Alcohol and Drug Abuse Prevention
New Hampshire Department of Health
  and Human Services
State Office Park South
105 Pleasant Street
Concord, New Hampshire 03301
TEL (603) 271-6104
FAX (603) 271-6116

NEW JERSEY

Mr. Terrence O'Connor
Assistant Commissioner
Division of Alcoholism, Drug Abuse,
  and Addiction Services
New Jersey Department of Health, CN 362
Trenton, New Jersey 08625-0362
TEL (609) 292-5760
FAX (609) 292-3816

NEW MEXICO

Ms. Lynn Brady
Director
Behavioral Health Services Division
New Mexico Department of Health
Harold Runnels Building, Room 3200 North
1190 St. Francis Street
Santa Fe, New Mexico 87501
TEL (505) 827-2601
FAX (505) 827-0097

NEW YORK

Ms. Jean Somers-Miller
Commissioner
New York State Office of Alcoholism
  and Substance Abuse Services
1450 Western Avenue
Albany, New York 12203-3526
TEL (518) 457-2061
FAX (518) 457-5474

NORTH CAROLINA

Julian F. Keith, M.D.
Director
Substance Abuse Services Section
Division of Mental Health, Developmental
  Disabilities, and Substance Abuse Services
North Carolina Department of
  Human Resources
325 North Salisbury Street
Raleigh, North Carolina 27603
TEL (919) 733-4670

NORTH DAKOTA

Mr. Don Wright
Unit Manager
Substance Abuse Services
Division of Mental Health and Substance
  Abuse Services
North Dakota Department of Human Services
Professional Building
600 South Second Street, Suite #1E
Bismarck, North Dakota 58504-5729
TEL (701) 328-8922/8920
FAX (701) 328-8969

OHIO

Ms. Luceille Fleming
Director
Ohio Department of Alcohol and
  Drug Addiction Services
280 North High Street
Two Nationwide Plaza, Fifteenth Floor
Columbus, Ohio 43215-2537
TEL (614) 466-3445
FAX (614) 752-8645

OKLAHOMA

Mr. Dennis Doyle
Deputy Commissioner
Substance Abuse Services
Oklahoma Department of Mental Health
  and Substance Abuse Services
1200 NE Thirteenth Street
Oklahoma City, OK 73117

Mailing Address

P.O. Box 53277, Capitol Station
Oklahoma City, Oklahoma 73152
TEL (405) 522-3858
FAX (405) 522-3650

OREGON

Ms. Barbara Cimaglio
Director
Office of Alcohol and Drug Abuse Programs
Oregon Department of Human Resources
Human Resources Building, Third Floor
500 Summer Street, NE
Salem, Oregon 97310-1016
TEL (503) 945-5763
FAX (503) 378-8467

PENNSYLVANIA

Mr. Gene Boyle
Director
Office of Drug and Alcohol Programs
Pennsylvania Department of Health

Mailing Address

P.O. Box 90, Room 933
Seventh and Forester Streets, Room 933
Harrisburg, Pennsylvania 17108
TEL (717) 787-9857
FAX (717) 772-6959

RHODE ISLAND

Sherry Knapp, Ph.D.,C.A.S., C.M.H.A.
Associate Director of Health
Division of Substance Abuse
Rhode Island Department of Health
Cannon Building, Suite 105
Three Capitol Hill
Providence, Rhode Island 02908-5097
TEL (401) 277-4680
FAX (401) 277-4688

SOUTH CAROLINA

Beverly G. Hamilton, M.Ed., M.H.A.
Director
South Carolina Department of Alcohol
  and Other Drug Abuse Services
3700 Forest Drive, Suite 300
Columbia, South Carolina 29204-4082
TEL (803) 734-9520
FAX (803) 734-9663

SOUTH DAKOTA

Mr. Gilbert Sudbeck
Director
Division of Alcohol and Drug Abuse
South Dakota Department of Human Services
Hillsview Plaza, East Highway 34
c/o 500 East Capitol
Pierre, South Dakota 57501-5070
TEL (605) 773-3123/5990
FAX (605) 773-5483

TENNESSEE

Stephanie W. Perry, M.D.
Assistant Commissioner
Bureau of Alcohol and Drug Abuse Services
Tennessee Department of Health
Cordell Hull Building, Third Floor
426 Fifth Avenue, North
Nashville, Tennessee 37247-4401
TEL (615) 741-1921
FAX (615) 532-2419

TEXAS

Ms. Terry Faye Blier
Executive Director
Texas Commission on Alcohol and Drug Abuse
9001 North IH 35
Austin, Texas 78753-5233
TEL (512) 349-6600
FAX (512) 837-0998

UTAH

Mr. Leon PoVey
Director
Division of Substance Abuse
Utah Department of Human Services
120 North 200 West, Room 413
Salt Lake City, Utah 84103
TEL (801) 538-3939
FAX (801) 538-4696

VERMONT

Mr. Tom Perras
Director
Office of Alcohol and Drug Abuse Programs
Vermont Agency of Human Services
108 Cherry Street
Burlington, Vermont 05402
TEL (802) 651-1550
FAX (802) 651-1573

VIRGINIA

Louis Gallant, Ph.D.
Acting Director Office of Substance Abuse
Services Virginia Department of Mental Health,
  Mental Retardation, and Substance
  Abuse Services
109 Governor Street

Mailing Address

P.O. Box 1797
Richmond, Virginia 23214
TEL (804) 786-3906
FAX (804) 371-0091

WASHINGTON

Mr. Kenneth D. Stark
Director
Division of Alcohol and Substance Abuse
Washington Department of Social and
  Health Services

FedEx ONLY

612 Woodland Square Loop, SE, Building C
Lacey, Washington 98503-5330

Mailing Address

P.O. Box 45330
Olympia, WA 98504-5330
TEL (360) 438-8200
FAX (360) 438-8078

WEST VIRGINIA

Mr. Jack Clohan, Jr.
Director
Division of Alcohol and Drug Abuse
Office of Behavioral Health Services
West Virginia Department of Health
  and Human Resources
Capitol Complex
1900 Kanawha Boulevard
Building 6, Room 738
Charleston, West Virginia 25305
TEL (304) 558-2276
FAX (304) 558-1008

WISCONSIN

Mr. Philip S. McCullough
Director
Bureau of Substance Abuse Services
Division of Supportive Living
Department of Health and Family Services
One West Wilson Street

Mailing Address

P.O. Box 7851
Madison, Wisconsin 53707-7851
TEL (608) 266-3719

WYOMING

Marilyn Patton, MSW
Deputy Director
Division of Behavioral Health
Department of Health
447 Hathaway Building
Cheyenne, Wyoming 82002
TEL (307) 777-6494
FAX (307) 777-5580

DISTRICT OF COLUMBIA

Mr. Jasper Ormond
Administrator
Addiction, Prevention, and Recovery
  Administration
1300 First Street, NE, Suite 300
Washington, DC 20002
TEL (202) 727-9393
FAX (202) 535-2028

TERRITORIES

PUERTO RICO

Mr. Jose Acevdo Martinez
Administrator
Puerto Rico Mental Health and Anti-addiction
  Services Administration
P.O. Box 21414
San Juan, Puerto Rico 00928-1414
TEL (787) 764-3795
FAX (787) 765-5895

VIRGIN ISLANDS

Carlos Ortiz, Ph.D.
Director
Division of Mental Health, Alcoholism,
  and Drug Dependency Services
U.S. Virgin Islands Department of Health
Charles Harwood Memorial Hospital
Christiansted, St. Croix, Virgin Islands 00820
TEL (809) 773-1311, ext 3013 or 773-1992
FAX (809) 773-7900

INDIAN NATION

RED LAKE BAND OF THE
CHIPPEWA INDIAN TRIBE
Mr. Bobby Whitefeather, Sr.
Chairman
Tribal Council
Red Lake Band of the Chippewa Indian Tribe
P.O. Box 574
Red Lake, Minnesota 56671
TEL (218) 679-3341
FAX (218) 679-3378

PACIFIC BASIN JURISDICTIONS

AMERICAN SAMOA
Mr. Faafetai I'aulualo
Chief
Division of Social Services
Department of Human Resources
Government of American Samoa
Pago Pago, American Samoa 96799
TEL (684) 633-2696
FAX (684) 633-7449

GUAM
Ms. Elena I. Scragg, MS, MHR, IMFT
Director
Department of Mental Health and
  Substance Abuse
Government of Guam
790 Governor Carlos G. Camacho Road
Tamuning, Guam 96911
TEL (671) 647-5445
FAX (671) 649-6948

COMMONWEALTH OF THE
NORTHERN MARIANA ISLANDS
Dr. Isamu Abraham
Secretary of Health
Department of Public Health Services
Commonwealth of the Northern Mariana Islands
P.O. Box 409 CK
Saipan, MP 96950
TEL (670) 234-8950 ext 2001
FAX (670) 234-8930

REPUBLIC OF PALAU
The Honorable Masao Ueda
Minister of Health
Ministry of Human Services
Palau National Hospital
Republic of Palau
P.O. Box 6027
Koror, Republic of Palau 96940-0504
TEL (680) 488-2813
FAX (680) 488-1211

REPUBLIC OF THE
MARSHALL ISLANDS
Mr. Donald Capelle
Secretary
Ministry of Health Services
P.O. Box 16
Majuro, Marshall Islands 96960
TEL (692) 625-3355/3399
FAX (692) 625-3432

FEDERATED STATES OF MICRONESIA
Eliuel K. Pretrick, MO, MPH
Secretary
Department of Health Services, FSM
P.O. Box PS 70
Palikir, Pohnpei FM 96941
TEL (691) 320-2619
FAX (691) 320-5263